Sweat

by Grace Hansen

BEGINNING SCIENCE:
GROSS BODY FUNTIONS

Abdo Kids Jumbo is an Imprint of Abdo Kids
abdobooks.com

abdobooks.com

Published by Abdo Kids, a division of ABDO, P.O. Box 398166, Minneapolis, Minnesota 55439.
Copyright © 2021 by Abdo Consulting Group, Inc. International copyrights reserved in all countries.
No part of this book may be reproduced in any form without written permission from the publisher.
Abdo Kids Jumbo™ is a trademark and logo of Abdo Kids.

Printed in the United States of America, North Mankato, Minnesota.

102020

012021

THIS BOOK CONTAINS
RECYCLED MATERIALS

Photo Credits: iStock, Science Source, Shutterstock

Production Contributors: Teddy Borth, Jennie Forsberg, Grace Hansen
Design Contributors: Dorothy Toth, Pakou Moua

Library of Congress Control Number: 2019956471

Publisher's Cataloging-in-Publication Data

Names: Hansen, Grace, author.

Title: Sweat / by Grace Hansen

Description: Minneapolis, Minnesota : Abdo Kids, 2021 | Series: Beginning science: gross body functions |
    Includes online resources and index.

Identifiers: ISBN 9781098202408 (lib. bdg.) | ISBN 9781644943878 (pbk.) | ISBN 9781098203382 (ebook)
    | ISBN 9781098203870 (Read-to-Me ebook)

Subjects: LCSH: Human body--Juvenile literature. | Perspiration--Juvenile literature. | Sweat glands—
    Juvenile literature. | Excretion--Juvenile literature. | Hygiene--Juvenile literature.

Classification: DDC 612--dc23

Table of Contents

Don't Sweat It!

Sweat might feel kind of gross.

But it has an important job to do!

5

Your body is an amazing machine. And that machine works best at about 98.6 degrees F (37°C).

When you move, play, or have a **fever**, your body warms up. When your body warms up, the **nervous system** sends messages to the brain. The brain wants the body to be cool and comfortable.

9

The **hypothalamus** is a very small part of the brain. But it plays many important roles in the body. One of those roles is controlling body temperature.

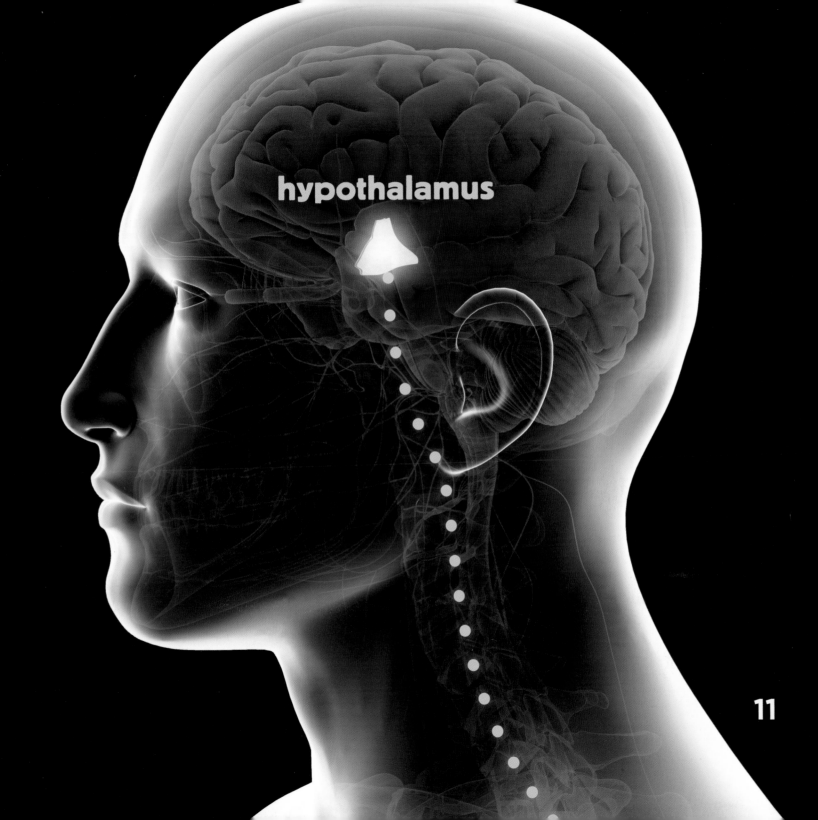

hypothalamus

11

The **hypothalamus** sends messages out to certain parts of the body. **Sweat glands** are one group that can receive a message.

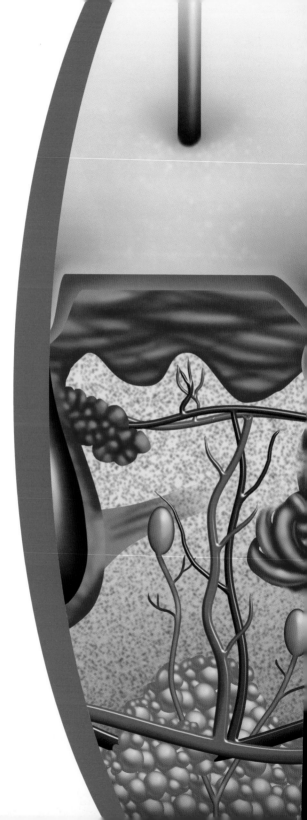

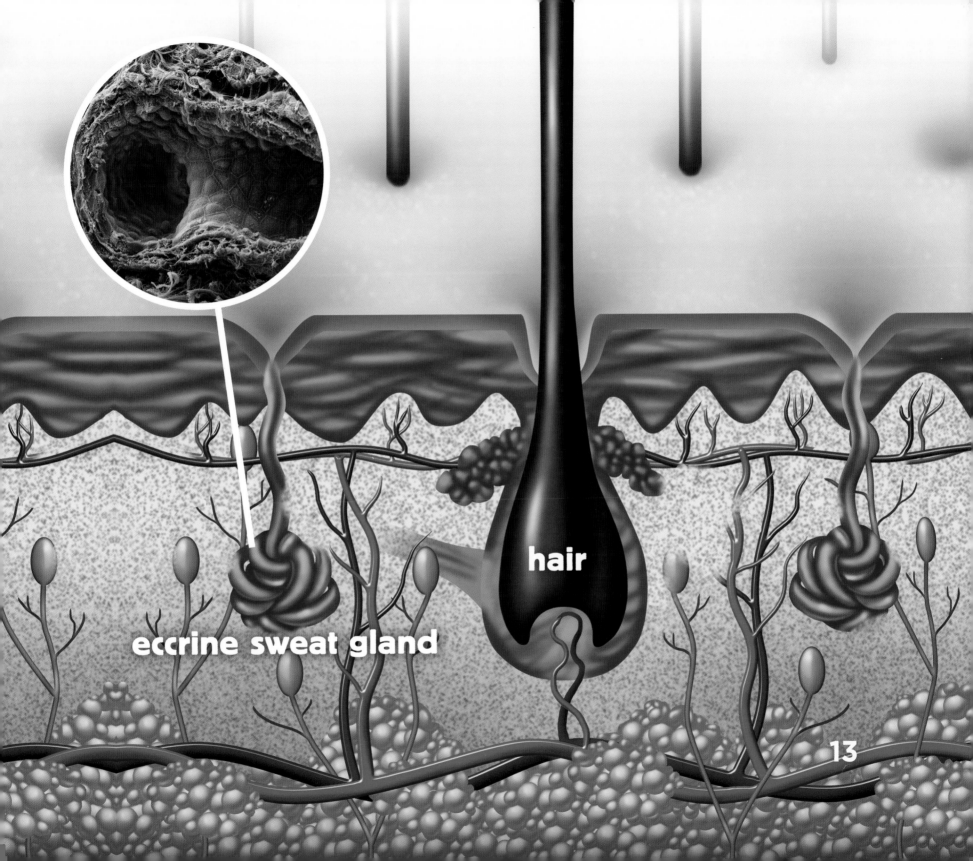

eccrine sweat gland

hair

13

Sweat Glands

There are two kinds of **sweat glands**. Apocrine sweat glands are in hairier areas, like the scalp and armpits.

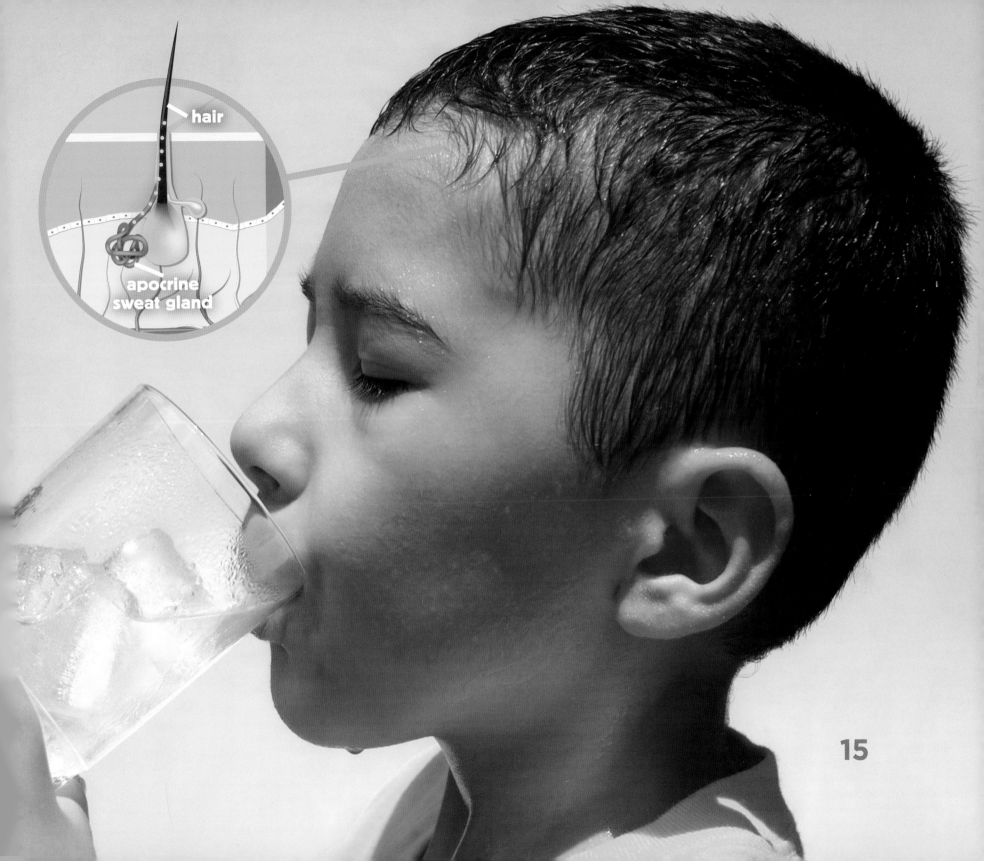

hair

apocrine sweat gland

15

Apocrine **sweat glands** release a fatty sweat. This sweat is broken down by bacteria on the skin. This can make it smell bad!

bacteria

Eccrine **sweat glands** cover most of the body. There are lots of them on the hands and feet. Sweat from these glands is less likely to cause an odor.

Chill Out

Sweat leaves glands through **pores**. Pores are tiny holes that cover your skin. The sweat then **evaporates** off of the skin. This cools your body down.

Let's Review!

- Sweat is natural and a very important thing that our body creates.

- Sweat is the body's response to being too hot.

- The brain likes the body to be at a certain temperature. When the body gets too hot, the **hypothalamus** tells the **sweat glands** to make sweat.

- The sweat makes it to the skin through pores and cools the body down.

- On a hot and dry day, sweat easily **evaporates** off of the skin. This helps cool the body even more.

Glossary

evaporate – to turn from a liquid to a gas.

fever – a body temperature higher than normal that is usually caused by the body's reaction to an illness.

hypothalamus – the part of the brain below the thalamus, important in the regulation of the autonomic nervous system and body temperature.

nervous system – a complex system of nerves and cells that carry messages to and from the brain and spinal cord to various parts of the body.

pores – tiny openings in the skin through which air, water, or sweat may pass.

sweat gland – a tiny gland located near the surface of the skin that makes and releases sweat.

Index

Abdo Kids
ONLINE
FREE! ONLINE MULTIMEDIA RESOURCES

Visit **abdokids.com**
to access crafts, games,
videos, and more!

Use Abdo Kids code
BSK2408
or scan this QR code!

24